3D Printing

How to Make Money Online Leveraging Technology with a 3D Printing Business

Table of Contents

Introduction

First and foremost I want to thank you for downloading the book, "3D Printing: How to Make Money Online Leveraging Technology with a 3D Printing Business".

In this book you will learn how to start your own 3D printing business and sell online with or without a 3D printer of your own.

Apparently, 3D printing technology is not really new but it has becoming more popular nowadays. Aside from huge 3D printing companies and manufacturers utilizing this technology, 3D printing is also a growing home-based market allowing individuals to create and provide relative service at the comforts of their home. The ideas are almost endless and this book will show you 3D printing trends.

You really don't need to purchase expensive 3D printers and materials. This book will guide you on how you can start-up with this business without breaking the bank. Sure there is a clear business case here so let us show you how.

Thanks again for downloading this book, I hope you enjoy it!

Chapter 1

3D Printing: How it Works

What most of us doesn't know is that the technology of 3D printing has been around since 1896 but was completely discovered until 1990. However it was popular back then in the world of manufacturing, architecture and engineering. The technology has come a long way in a very short span of time. Charles Hull initially built it in the 1980s for making basic object made from polymer. Today, 3D printing is serving a myriad of companies and manufacturing sectors. From car components to aircraft to prosthesis and human organs, indeed 3D printing is very in demand. 3D printing services are also getting the hype being a cost-effective, environment-friendly and efficient method of manufacturing.

3D Printing was even called desktop fabrication before at is can form materials obtained as a powder. To create an object you will 3D digital model by drawing it on a computer with CAD software, from downloadable images in the internet or by scanning set of 3D images using 3D scanner. Afterwards you can start the process of printing of your three-dimensional object using special equipment.

The process of printing a three-dimensional object from a digital file is called additive process wherein successive layers of materials are being lay down until the whole object is finish. Such layers could be seen as very thinly sliced cross-sections of the created object.

3D scanners are using various technologies in creating a 3D model like volumetric scanning, time-of-flight, modulated light and structured light. It companies has also enabled their software to have 3D scanning features such as Google and Microsoft. It is a clear indication that future gadgets and devices could have a built-in 3D scanner making it easier for people to digitize real objects.

3D Printing Applications

3D printing has been considered essential in the medical industry since surgeons and health experts are able to create mock-ups of body parts of their patients to determine how to operate or treat them efficiently. Likewise it allows developers and designers to come up with better designs and technological innovations.

Toy makers, aerospace components, fabrications and other industries are now using 3D printers and are taking the advantage of 3D technology. Likewise it can give great amount of savings in terms of assembled products since it can print the exact products. Companies can also now experiment with new business ideas, create fresh designs; make alterations with their existing products without spending too much time and resources.

Indeed 3D printing technology has made production and manufacturing extremely convenient. It has impacted various industries such as automotive, consumer-based industries, product manufacturing, housing and real estates, industrial and business equipment, medical tools, architecture and education.

Software for Beginners

Not all of us are fond of working with complex software thus, we are giving you the simplest 3D modeling software you can try and download for free to get you started.

Google SketchUp – is free and fun to use. You can draw faces and edges with some basic tools and features that you can learn in a short span of time. This works better with Google Earth making it easier for you to build models that Google Earth can detect.

Blender – is a free open source, available for all operating systems. It is an in-house application and a powerful program which contains features similar with that of high-end 3D software.

3Dtin- is the simplest among these software programs allowing you to draw straight from your browser.

OpenSCAD- allows you to create solid 3D CAD designs and is free. It is compatible with Windows, Linux and Mac OS X however; it focuses on the CAD aspects instead of 3D modeling aspects.

Although you don't have any experience in terms of 3D model and designing, you can actually learn how to do it with the help of these software and other applications online such as Rhino. This will take you weeks to master the modeling tools while to become a professional user, you have to study and practice them for at least half a year.

Starting Your Own 3D Printing Business

Apparently 3D printing has becoming a more popular trend nowadays across the globe. Entrepreneurs are talking about it and see huge potential in this kind of business. Regardless if you are still a student, a stay-at-home mom, a job seeker or someone who wants to earn additional income, you can learn and earn with 3D printing.

Let this book show you how it is actually easier and manageable to set-up a 3D printing business without investing millions. We will discuss more of the steps to follow and the strategies to use to make this business a success.

For small businesses and new entrepreneurs, the greatest challenge would be cost of technology to turn their ideas into a tangible item. And because even the thought of creating a prototype is overwhelming, these ideas were left behind. This is why 3D printing technology is considered a game changer by most manufacturers and companies. For once, creating and producing tangible products cost huge money, 3D printing made it possible for people to do prototypes and to mass produce with the fraction of cost. With just a machine, 3D models and perseverance, one can make a three-dimensional object of almost anything on the spot.

Chapter 2

Printing Ideas

Apparently the possibilities are endless and overwhelming as what you could do with 3D printing. As a newbie in this business, you will definitely have plenty of question in mind; where to start? What tools to learn? How could I equipment without paying too much? What certain product or service do I need to offer?

With all these questions in mind, one would feel extremely excited on how 3D printing can fuel their creativity and become a successful entrepreneur. Likewise you will realize that you can actually earn money with 3D printing and at the same time, get to enjoy your interest and hobbies. Below are some of the tried and tested ways on making money out of 3D printing.

Create and Sell

This is the most inexpensive and simplest way to get started. It is as if you are creating an application and sell it for royalties.

- Learn how to use CAD software with free CAD tools online. Beginners could easily create a complex-looking design after basic tutorials. Professional and sophisticated designs can be achieved using these tools. Leveraging with the online tutorials available for free as well as engaging on related communities would really help you to jumpstart your 3D printing business.

- Take advantage of the 3D printing technology by developing intricate designs and shapes easily. You can also offer customize designs for business establishments, companies and other institutions at very affordable price. There are hundreds of thousands of designs online for your inspiration.

- After completing your designs through CAD, make sure it is water tight meaning, its 3D printable. There are free tools online that will help you correct errors and enhance your creations.

- You can sell you ready-to-print digital design through various online markets such as Ebay, Amazon or through your website. You can also sign-up for free sites where clients are looking for printable 3D designs or can request for a custom-made one.

Offering 3D Printing Services

While this idea seems simple, it is not quite as easy. One should invest at least hundreds of thousands to purchase a good quality printer.

- Acquire a printer that is cost-effective and can provide you high value output, versatile to use and user-friendly.

- Once purchased, take time to master it and learn the basic operation, troubleshooting and other must-know about the printer. Factors such as travel speed, layer height and extrusion can greatly affect the output. The proper feedstock should be use depending on the type of materials. The G-code or the software use to convert the printable file into a machine code is essential. The whole process could be fun and overwhelming for you but with proper research and studying, you can soon operate your printer efficiently.

- If you are not really that creative and is more on the selling and promoting side, better hire a good artist or designer. Once you are sure of the work flow and confident enough of the output, you can join the community of 3D printing networks. It enables business person like you to accept orders from customers with pre-made designs by listing your printer.

Come-up with other Services or Products that Leverages 3D Printing

Some entrepreneurs come up with an online service or product using 3D printing. However this requires creativity and money but is considered the highest profitable business idea among the three. Here you are running on online business, selling services or products created by 3D printing and other related services.

- A valuable service in terms of 3D printing is relative associated with the people's freedom to design and create. One can easily modify designs and customized products according to the client's requirements. Everyone can create their product design and let 3D printing enhance it.

- After crafting a service or product to sell, it is easier for you now to create your website or sign-up for online markets. Focus on a certain feature of 3D printing to leverage. And since you are starting, you still may be in the process of acquiring professional grade printers and other tools to be used. You can consider partnering first with an already established 3D printing service provider. These establishments will print for you for certain fees.

To lessen the burden of shipping and billing processes, you have to carefully develop an efficient operational process for a smooth transaction. From clients' placing orders and sending you design, to printing of orders and shipping them to the client, you will need to be organized and systematic.

Chapter 3

Setting Up Your 3D Enterprise

Apparently, it is the 3D printing era. New businesses are arising taking advantage of 3D printing technology. It enables us to make our own designs and customized our own products. You can have it printed at home or through a 3D printing service company and sell it online. With minimal efforts, you can come up with countless products and various materials.

Below are easy steps to start up with your 3D printing enterprise and become a successful entrepreneur leveraging from this technology.

Purchase a 3D printer- the first and most important thing to acquire is a 3D printer. If you are planning to have your own printer instead of paying for printing services, make sure to choose the model that suits you, the nature of your printing business and is cost effective.

3D printers are different from the common printers you see as a 3D printer is able to print objects in a three dimension by printing layer by layer. The process is known as rapid prototyping. Depending on the complexity of the model, it may take you hours or even days to fully print a 3D model.

Fortunately there are simplified 3D printers available now for home use which is cheaper than the standard 3D printers. The materials to be use are also less expensive however; the outputs are not as accurate compared to the outputs of high grade 3D printer.

3D Printer Parts - below are the common printer parts you should have to be familiarized with.

- **Print bed** refers to where the printed parts will adhere during the production. The temperature can be heated or ambient. The non-heated beds are usually covered with painter's tape where the printed materials rest while the heated beds keep the parts warm to prevent materials from warping. The heated bed normally ranges from 40°C to 110°C throughout the printing process depending on the materials used. Be careful not to touch the insides of 3D printers while in used as it can cause blisters and skin burn.

- The **Extruder** is where you will feed the plastic filament to the hot-end. It pushes the filament through a stiff tube going to the hot end. There are printers with 2 extruders allowing the user to print different products simultaneously. However such versatility comes with more expensive cost and complexity as it requires extra hot-end and other parts. The Ultimaker 3D printer can be upgraded to have multiple extruders.

- The **Hot-End** consists of a temperature sensor and a heater as well as an extrusion end where the plastic filament will be pushed. It is usually assembled with aluminum block or modified in a barrel-type shape. Adjust your printers properly as the interface between the extruder and the hot-end can cause problems when not properly adjusted. Likewise the smaller the nozzle of the hot-end is, the more detailed the printed product becomes but this also take longer.

- **Plastic filament** serves as the ink. If an inkjet printer requires ink to produce designs, 3D printers use plastic filament. Take note that some materials can stress the printer as not all of them are compatible to use.

Choose your software – as mentioned earlier, you have to choose which software to use in creating designs. Good thing is, due to the popularity of 3D printing nowadays, the software is getting more affordable and user-friendly. There are plenty of open sources or free software like OpenSCAD, Wings3F, SketchUp and Blender.

One just needs to start practicing on their designing skills. When your business is already established, you may consider investing more for advanced software.

Conduct your research – do your homework and make research on your target market and determine the products that are in demand. Find out if your designs and ideas are profitable and sellable. Do you prefer doing a specific niche or would want to produce goods with high demands? You must also find out if your designs still make it to trends' list or do you need to offer fresh designs or have variations of the current one. Lastly, learn the current price of the similar service or products you are to offer.

The materials to use – you also have to invest in materials to use thus, make sure that you are getting quality products but are still within your budget. There are various materials to be use in 3D printing which makes it harder to decide. Below are some of the most common proprietary materials to guide you.

PLASTICS

- Nylon or Polyamide) are typically used in powder form with the FDM or filament form. These are flexible, strong and durable plastic materials reliable for 3D printing. You can also combine this material with powdered aluminium to create another material for sintering called Alumide.

- Another common plastic used is ABS which is widely used in filament form. It comes in various colors and is a strong plastic which makes it very popular.

- Another common plastic material is called PLA. It is bio-degradable plastic and can be used in resin format as well as filament. It also comes in different colors which is very useful in 3D application.

- LayWood is a special 3D printing material developed in filament form and is a polymer/wood composite.

Metals

Aside from plastics, various metals and metal composites are also widely used for 3D printing. The most common are cobalt and aluminium derivatives. Stainless Steel is also popular since it can be plated with other types of materials to achieve a bronze or gold effect. Silver and Gold also make it to the list since you can directly print on them which is the typical application used in jewellery sector. Likewise titanium is another strong metal material available in powder form.

Ceramics

It is relatively, a new material used in 3D printing. It has to undergo similar processes using traditional methods such as glazing and firing during post-printing.

Paper

The standard A4 paper is another common material which is considered a cost-effective one that can obtained locally. Likewise using paper in printing 3D models is safe, recyclable, and environment-friendly and does not require post-processing.

Food

Apparently food substances became a part of the experiments for 3D printing. Chocolate bars and hard candies are the common products used. Still on-going researches aim to do the same with meat and pasta and to come-up with balanced whole meals using 3D printing technology.

Practice – once you have the materials to use and you already know how to produce them, try and practice it. Use different designs and various materials to find out which of the materials created are of best-price quality. This could take time but it is ideal to spend over materials and processes that best suits you. Moreover you will need to improve your crafting and designing skills.

Get ready to sell – after the products and services to offer are being tried and tested and you think everything is set, you are good to go. You have to produce more and start selling. Whilst selling locally is recommended, selling your products online is more profitable since a wider range of audience will have access on your products.

Those are the basic steps in getting set-up and start your 3D printing enterprise. Continuously, research on new designs and materials, practice more on them and advertise your business well. There are online communities to help you. You can also read blogs and self-help books for more reference with 3D printing.

Knowing your tools and materials to use is very essential. If possible download and print all the necessary upgrades for your printer and acquire replacement parts. Start printing small items such as an ear bud, a small cartoon design or a key. Choose a design that lots of people are doing and using so you can compare your own work with their products. Likewise you will have a better comparison and basis for fixing possible problems to maximize the print quality of your printer.

Moreover, 3D printers vary in terms of successful print. By learning them you can easily identify problems and can troubleshoot it.

Chapter 4

Understanding Basic 3D Printing Process

Regardless of the approach of the 3D printer to use, the printing process is basically the same. Below are the generic steps to follow.

Step1: Produce a 3D Model

With the aid of software such as CAD software, produce or create a 3D model depending on your choice or the client's requirements. Some software can give you hint regarding the structural integrity of the finished product. You can also create a virtual simulation on certain materials you want to use for your product and determine how it will behave under specific conditions.

The OpenSCAD program allows users to draw their desired product by writing codes instead of traditional drawings. 3D models can also be generated from videos or photos by using certain software such as Autodesk's 123D Catch.

Step 2: Convert to STL

STL is the acronym for standard tessellation language. It is a file format specifically developed for 3D systems for use by SLA machines or stereolithography apparatus. Most 3D printers today use STL files and other proprietary files.

The STL files are the foundation of the 3D printed products. Exporting a 3D model as an STL file means being able to have the model sliced and printed.

Step 3: Transferring to AM Machine/ STL file manipulation

Next, you will copy the STL file to your computer where the 3D printer is connected. Choose the orientation and size for printing.

Step 4: Setting Up the Machine

Every machine varies in terms of usage and requirements to print 3D objects. This includes refilling of binders, polymers and other materials the printer will use.

Step 5: Build

Mostly, the building process is automatic. The layer could be thicker than 0.1 mm depending on the size of the project. Likewise the process can take hours or days to finish depending on the machine, materials and design to print so make sure to check the machine often to prevent errors.

Step 6: Post-Processing

After removing the printed object from the machine, you can start with post-processing. Wear gloves when removing them to protect you from toxic chemicals and hot surfaces.

Some printers would require processes such as brushing-off remaining powder or washing the printed object to remove other elements. Before actually doing this, be sure that the object is already stable since some materials tend to be week after printing. Take extra caution to avoid breakage.

Choosing a 3D Printer

The costs to establish a 3D printing business varies. The commercial printers intended for creating prosthetics or mechanical parts for example, normally costs $20,000-$70,000 while personal or home printers would cost you $2,000. This could be a huge money on your part but a very manageable startup cost for a great business.

Most start-up 3D printers are amenable to modification to fit your needs. Moreover the software to use are mostly open source but you can also acquire modelers and commercial slicers commonly used in 3D printing.

One of the most common type of 3D printers is called the additive manufacturing type. It creates parts by adding materials which is very useful in the field of manufacturing. The 3 approaches to additive manufacturing are photopolymerization, granular materials binding and molten polymer. Each has their own features;

Photopolymerization - this uses light to turn liquid materials to solid products.

Granular materials binding - use hot air, lasers and other energy sources to bind layers of powder to the desired shape.

Molten polymer - or the MPD extrude molten materials to create the desired product.

Buying Options

Turnkey, DIY or kit? There are several 3D printing solutions that are quite affordable and goes well with your budget. There are ready-to-print device such as the Up 3D printer that only requires little maintenance while the MarketBot Replicator needs more maintenance but is more versatile compare to Up 3D printer.

Similarly, there are printers that are seemingly cheap but require filament cartridges that normally cost up to three times the going market rates.

For the kits, the Ultimaker and Printrbot are the best examples. You can assemble them using a manual and the assembly is more on its mechanical construction and are ready-to-go.

The Rostock and RepRap printers are **DIY 3D printers** which are a combination of various wood bits or metals, electronics and 3D-printed parts. The pre-printed parts can be ordered from sources on online marketplaces such as eBay while the mechanical or electronic parts are available as well.

DIY 3D and kit printers are those who likes tinkering and are ready to get their hands dirty. Whilst these printers are difficult to assemble for beginners, these doesn't compromise the quality of prints. However these types of 3D printers require modifications and upgrades more often. If you choose DIY and kits, keep in mind that it is a long term investment so expect the challenges and be accustomed to quirks.

The Power of 3D Printing

With the numerous scams online, making money using the internet is quite difficult nowadays. It requires patience and hard work especially if you are offering products and commodities. Others would have invested years ago that is why they are now making millions of dollars.

Unlike those investment scams and other money-making ways offered online, 3D printing business is the authentic business you can run online. While you are offering digital tools, you are also able to provide your customers physical products. You can create personalized iPad case, iPhone cases and relative products and be able to show it to your customers. Hence they will have an idea on how you are doing the items intended for them which will gain their trust over your service.

Although it could be a tedious task at the beginning, 3D printing business is a sure way to gain profits. Creating products for different niches will allow you to have access over billions of people who are using such products or services.

Chapter 5

Business Opportunities

From toys to life-saving products, there are various business opportunities and trends that do not existed before, thanks to 3D printing technology. The advent of this technology has paved way to quick designing and manufacturing of products and commodities. As this technology become more accessible to the public, it has also redefining the people's perception about 3D imaging and manufacturing and the way entrepreneurs are running their businesses.

Below are some services and products that leverage 3D printing technology.

GADGET ACCESORIES - apparently, billions of people are now into gadgets hence, this business trend is booming in the recent years. For an instance, ear buds, ear plugs and smart phone cases can now be personalized. 3D printed ear buds are becoming popular since most of us are fed-up with the generic sizes ear buds. It also allows customers to create their own design and choose the color of their headphones.

Phone cases can be designed with the customer's name or their preferred logo and color as well. Small online businesses saw opportunities every time there is a new smart phone being launched in the market.

SOUVENIRS - ceramic figurines, photos or key chains are the usual souvenirs we encounter but 3D printing can make your souvenirs more unique and personalized. There are companies now offering awesome souvenirs such as 3D printed fetuses. Instead of the regular ultrasounds, they generate the required daa using ultrasound images or newborn photos to have it printed into figurines.

EDIBLE PRODUCTS - another brilliant innovation is using 3D printing technology in edible goods. Chocolates and candy bars are among the most popular. Company logos, names or even portraits are unique images that can be printed on your candy or chocolate bars. Other companies are making this trend as well in pasta and other food products.

OTHER CREATIVE PRODUCTS - whatever you can imagine, 3D printing can provide it to you. From toys to jewelries, to make-up kits and home decors, customers can now connect to 3D designers through online marketplaces. They can upload their chosen designs and products for production.

Companies like Crayon Creatures can also help your little kids by bringing their art and drawings into reality. You just have to send them the drawing and they will create a three-dimensional copy of it. Plush toys and baby photos are also a huge hit since it can be 3D printed for souvenirs and keepsakes.

Ideas are almost everywhere thus the challenge for entrepreneurs like you is to come up with the most unique product that customers will buy. 3D printing is an efficient way to test your ideas.

You can test print jewelries and other fashion accessories before mass producing them. You can also customized mugs, pens and other promotional items, holiday gifts, giveaways and marketing collaterals. Likewise you can specialize in machine parts.

Sell Your Items Online

Manufacturers know how 3D printing can be utilized to increase their profits and gain more customers especially small businesses. The possibilities of 3D prints are endless and the internet makes it more manageable to handle a 3D printing business.

Our concept of internet as a tool for communication and disseminating information have changed. Internet also serves as a huge marketplace today and everyone seems like doing business online. The advent of 3D printing technology and its power to create things from imagination out of people's homes paved way to the concept of making it a home-based business.

And because not all of us can make our websites or has the resources to to run a physical store, selling platforms such as Amazon, Etsy and Ebay are a huge help. These platforms are popular nowadays which can help sellers reach millions of customers online.

Quick Turnaround and Personalization

And because 3D printing technology can make all these objects and services possible, the need for 3D printing services is higher making 3D printing less costly which is good for beginners in the business. You can provide your customers quick turnaround of products. You might also be able to provide simple spare parts, cases for gadgets and smart phones and plastic toys.

Apparently, customers want their items delivered fast making 3D printing a unique and preferred method of manufacturing products. This technology will allow you yo consistently deliver goods in a very short span of time meeting your customers' requirements and deadlines.

Moreover, you are also able to give a more personalized product for your customers incorporating all of their wants and needs in a single item.

Chapter 6

How to Start without a 3D Printer

Don't worry if you do not have a 3D printer, you can still start your 3D printing business and enjoy the benefits of it. There are various ways on how you can turn your 3D models and digital files to a physical object. Nowadays various printing service providers can print for you using a wide range of materials and printers depending on your requirements.

And even if you do have a 3D printer, other commercial printing companies can make your plastic prototype printed using materials that ordinary 3D printers cannot use such as stainless steel or titanium. Below are some of the most popular 3D printing service providers.

PONOKO - is a company offering a wide range of materials from ceramics to plastics, to gold plate and stainless steel. They also have CNC routing and laser cutting in various materials so you can sustain your 3D printed project with more custom parts. Customers agreed that their prints are of good quality and cost-efficient with a good support system. Currently Ponoko operates in several regions.

SHAPEWAYS- another popular 3D printing service provider, Shapeways caters to professional designers and hobbyists. They provide high quality prints and uses a wide range of materials such as sterling silver. Although they are offering the lowest prices, the shipping times are inaccurate since they are operating in the

Netherlands. However they are now building production facility in New York for their U.S-based customers.

Moreover Shapeways has a marketplace allowing other users to have their own online shops similar with eBay which is helpful for beginners.

SCULPTEO - is a France-based company which offers materials such as resin, ceramics, multicolored plastics and silver-plated plastic. They are selling sample kits of their materials and offers preview of printability which is useful for customers. They are also providing "creator" apps where you can create designs right away. Usually they ship printed products for a couple of days to a month depending on the materials used. Sculpteo's website also allows customers to post their designs and products for sale.

I.MATERIALIZE- is based from outside the Netherlands with a user-friendly interface. They are offering more than 20 different materials and can prints huge objects. Customers can also sell their creations on the company's gallery.

REDEYE - serves engineering, medical, architectural and aerospace industries. They print objects in UV photopolymers, thermoplastics and resins. And because they are providing professional services, their prices are higher compared to the above printing service providers.

ZOOMRP.COM - is a self-service website catering to those who need rapid prototyping. They specializes in same-day printing and shipping or for a week which made them costly. This provider best suited those who are knowledgeable with 3D design and CAD operations.

Manufacturers and entrepreneurs know that 3D printing has a lot of potential and it is starting to get more hyped these days. Before, the machines to be used seem to be a big deal but thanks to the advent of more approachable and cheaper 3D printers, even an average joe can acquire them. These printers may not produce high-end quality products but still, the technology is imposing major breakthroughs that are beyond plastic wares.

With the help of these huge 3D printing service providers, we now have the option to create almost everything. Nonetheless it could still be intimidating to use especially if a person lacks the knowledge on how it works. Hence, equipping with yourself with the basics of 3D printing will help you once you are decided to join the fanfare of 3D printing business.

Nonetheless we need not to own 3D printers or necessary materials to benefit from this technology. Innovative companies such as Shapeways make it possible for us to gain profits and enjoy 3D printing. One great thing is that you don't also need to take care of the inventory. Anytime you place an order to these 3D printing service providers, they will take care of them and ship them to you.

Chapter 7

More Tips for Starters

To be honest, 3D printing can be tricky and overwhelming. Likewise home 3D printing technology still needs more improvement and beginners who brought their 3D printers will face multiple challenges at the onset. One should need to learn more to achieve reliable and high quality 3D print results.

An expert guidance will extremely help that's why we gathered tried and tested tips and advice from 3D printing experts.

- Despite the hype of various 3D printer brands, understand that home 3D printing is not really as simple as 1,2,3. People should first know how their printers work. Even kits and DIY 3D printers are tricky and only few of them are really reliable. Even if you are lucky enough to work through your printer, you will still encounter future problems and issues about it. So be patient in reading that manual and guides and be sure to ask questions to the selling staff upon purchasing your 3D printer.

- Do not rely on the price as this is not an accurate indicator on how efficient your 3D printer can be. Make sure that your printer is properly tuned before using it. Check all its parts, from the bed, belts to screws and pulleys. Recognizing the problem could be hard at first as issues with 3D printers are often not obvious. Fortunately forums and online supports can help you with it. Practice is also a must and eventually, you will learn what seems to be wrong with your 3D printer just by the way it sounds.

- The first few layers of print are the most important part hence, calibrating your printer is necessary. Building a good solid foundation for your 3D print to have a better result. When something goes wrong at the onset, cancel printing immediately as it is not going to get better.

- Use cheats and tricks you have learn in 3D printing. The goal here is to have a complete set of objects printed so make sure that the prints stay in the bed. Moreover, do not print too many objects at a time or it will the ruin the rest of the prints. Test print first and use slicer for complicated 3D models.

- Like all other equipment and machineries, keeping your 3D printers clean is the key to longevity. This also ensures quality prints and extended lifespan. Regularly clean old adhesives to keep the prints properly stick. Likewise 3D printer could be easily affected by its environment, the temperature and your mood. Avoid printing plastic in cold areas. Adjust the temperature according to the filaments to use. Have a clear working area and always be organized.

- Make sure that the model is watertight or completely closed. Not doing this before exporting it to STL file will make it unsuitable for printing with gaps and holes on it.

- One important part of 3D printing is slicing since it turns digital 3D model into an information that 3D printers understood. Get a slicer software to come up with the most precise toolpath information and achieve best printing results. Slicing is an essential part on the 3D printing process thus, you should find the best slicer as much as you need a quality 3D printer.

- Just because you know how to use OpenSCAD, CAD and other software, this means you will be able to produce the best results. Do some research to avoid mismatch or products or errors once they are printed. Consider factors such as wall thickness, the best printing orientation and the materials to use, scale and sizing as these greatly affect the output.

Another proven tip, one must learn from the mistakes of other people. Although you exactly know the basics of 3D printing and have the idea how to run a 3D printing business, chances are you will be facing problems in the future. From trouble-shooting of your printer to achieving the right quality of product, you will need other peoples' advice and support.

Take it from those who failed and succeed in this kind of business. They know better in terms of making your items accessible for more people, the right marketing strategy and other related issues.

Moreover, the most important part of all these is business planning. Regardless how big or small your 3D printing business is, it requires planning on your part. Before diving more into the details of the printer and materials to use, creating and planning your business is an essential step to start.

Conclusion

Thank you again for downloading this book!

I hope this book was able to help you to understand how 3D printing works and how you can actually gain profits from it.

The next step upon successful completion of this book is to get started with your online 3D printing business. Indeed, this technology encourages and fuels innovation giving people the freedom to create and produce objects with lesser tools an reduced cost.

Nonetheless, 3D printing technology is an energy-efficient one that we should take advantage of. Do not hesitate to ask and learn more from other experts or online communities. Nurture the knowledge and be a successful entrepreneur with the help of 3D printing technology.

Finally, if you enjoyed this book, please take the time to share your thoughts and post a review on Amazon. It'd be greatly appreciated!

Thank you and good luck!

www.ingramcontent.com/pod-product-compliance
Lightning Source LLC
Chambersburg PA
CBHW020958180526
45163CB00006B/2416